I0882907

A Maker's Guide to

AXLES

Written by JOHN WOOD

Illustrated by AMY LI

LERNER PUBLICATIONS ◆ MINNEAPOLIS

First American edition published in 2021 by Lerner Publishing Group, Inc.

© 2021 Booklife Publishing

This edition is published by arrangement with Booklife Publishing. All rights reserved.

Editor: Madeline Tyler
Design: Amy Li

All rights reserved. International copyright secured. No part of this book may be reproduced, stored in a retrieval system, or transmitted in any form or by any means—electronic, mechanical, photocopying, recording, or otherwise—without the prior written permission or Lerner Publishing Group, Inc., except for the inclusion of brief quotations in an acknowledged review.

Lerner Publications Company
An imprint of Lerner Publishing Group, Inc.
241 First Avenue North
Minneapolis, MN 55401 USA

For reading levels and more information, look up this title at www.lernerbooks.com.

Main body text set in VAG Rounded Std. Typeface provided by Adobe Systems.

Library of Congress Cataloging-in-Publication Data

Names: Wood, John, 1990– author. | Li, Amy, illustrator.
Title: A maker's guide to axles / written by John Wood ; illustrated by Amy Li.
Description: First American edition. | Minneapolis : Lerner Publications, 2021. | Series: A maker's guide to simple machines | Includes bibliographical references and index. | Audience: Ages 6–10 | Audience: Grades 2–3 | Summary: "What is an axle, and how does it make life easier for us Earthlings? Join the Maker, a supersmart alien engineer, in a deep dive into axles, from wind turbines to rolling pins"—Provided by publisher.
Identifiers: LCCN 2020053480 (print) | LCCN 2020053481 (ebook) | ISBN 9781728416410 (library binding) | ISBN 9781728438306 (paperback) | ISBN 9781728418865 (ebook)
Subjects: LCSH: Wheels—Juvenile literature. | Axles—Juvenile literature.
Classification: LCC TJ181.5 .W66 2021 (print) | LCC TJ181.5 (ebook) | DDC 621.8/2—dc23

LC record available at https://lccn.loc.gov/2020053480
LC ebook record available at https://lccn.loc.gov/2020053481

Manufactured in the United States of America
1-48888-49202-12/7/2020

Table of CONTENTS

Page 4	Be a Maker
Page 6	Axles
Page 10	Parts of an Axle
Page 12	How an Axle Works
Page 14	Gears
Page 16	Big Gear, Little Gear
Page 18	Build a Spinning Toy
Page 24	Glossary and Index

Words that look like <u>this</u> can be found in the glossary on page 24.

BE A MAKER

Oh no! Sorry about this mess—Maker can be a bit clumsy sometimes. Say sorry, Maker.

Believe it or not, Maker is very smart when it comes to machines.

A machine is an object that makes a job easier to do. Maker wants to teach you about one of the simplest types of machine: an axle.

A watch is a machine. Even a pizza cutter is a simple machine.

AXLES

An axle is a thin, stiff rod that goes through the middle of a bigger, circular object called a wheel. The axle lets the wheel spin around.

Any wheel on a vehicle, such as a car or bike, has an axle.

Some axles are fixed to the wheels—when they turn, the wheel spins. Other axles stay still while the wheel spins around them.

Axles are everywhere. Here are a few examples.

In a wind turbine, the blades are the wheel, and the axle is in the middle.

A well has an axle. When it turns, it raises and lowers the bucket.

There are some surprising axles on this page!

The handle of a faucet is a wheel, and it is fixed to an axle. Turning the axle lets water flow out.

In a screwdriver, the handle is the wheel, and the metal part is the axle.

There is an axle in the middle of some rolling pins.

PARTS OF AN AXLE

In a wheel and axle machine, there needs to be something to make it move. This might be a person, such as someone pedaling a bike. It could also be an electric motor.

Axles also have bearings. Bearings are parts that go between the axle and wheel. They are used to take away friction, and they help the wheel or axle to spin smoothly.

Many bearings are made up of small metal balls. These are called ball bearings.

HOW AN AXLE WORKS

When a small, fixed axle is turned, it can turn a much bigger wheel. Axles let us spin wheels more quickly and easily.

Axles and wheels can spin either clockwise or counterclockwise.

Axles let us build machines that work harder, with wheels and gears that go faster. Wait, what are gears?

Maker says he is going to explain what gears are on the next page.

GEARS

A gear is a wheel that has teeth around the outside edge. The teeth help gears to fit together. When a gear turns around an axle, it turns other gears that are connected to it.

Machines, such as clocks, can be made up of lots of gears.

A gear will always spin in the opposite direction of the gear turning it. This means that if the first gear is spinning clockwise, the second gear will spin counterclockwise.

Gears aren't always facing the same way. Gears can fit together in all sorts of ways.

BIG GEAR, LITTLE GEAR

When gears of different sizes are put together, they can change the speed and power of machines.

Turning a big gear once will make small gears turn lots of times. This will turn the axle more, which gives you more speed.

Turning a little gear will make the big gear turn more slowly. However, this uses less force, so it is easier to turn.

Bicycles use different gears to help you pedal. It's easier to pedal in low gears because you're turning a small gear. Less force is needed to spin the wheels, but you will move more slowly.

Different sizes of gears on a bicycle

It's harder to pedal in high gears because you're turning a big gear. More force is needed, but you can be speedy.

BUILD A SPINNING TOY

It is time to build! We will be using axles to make a toy. It will have a part that spins when you turn the axle.

YOU WILL NEED:

- Glue
- 2 bottlecaps
- 3 long, thin wooden sticks
- Scissors
- Colored pencils
- A straw
- A small cardboard box that is shorter and thinner than the sticks of wood
- A few small pieces of cardboard

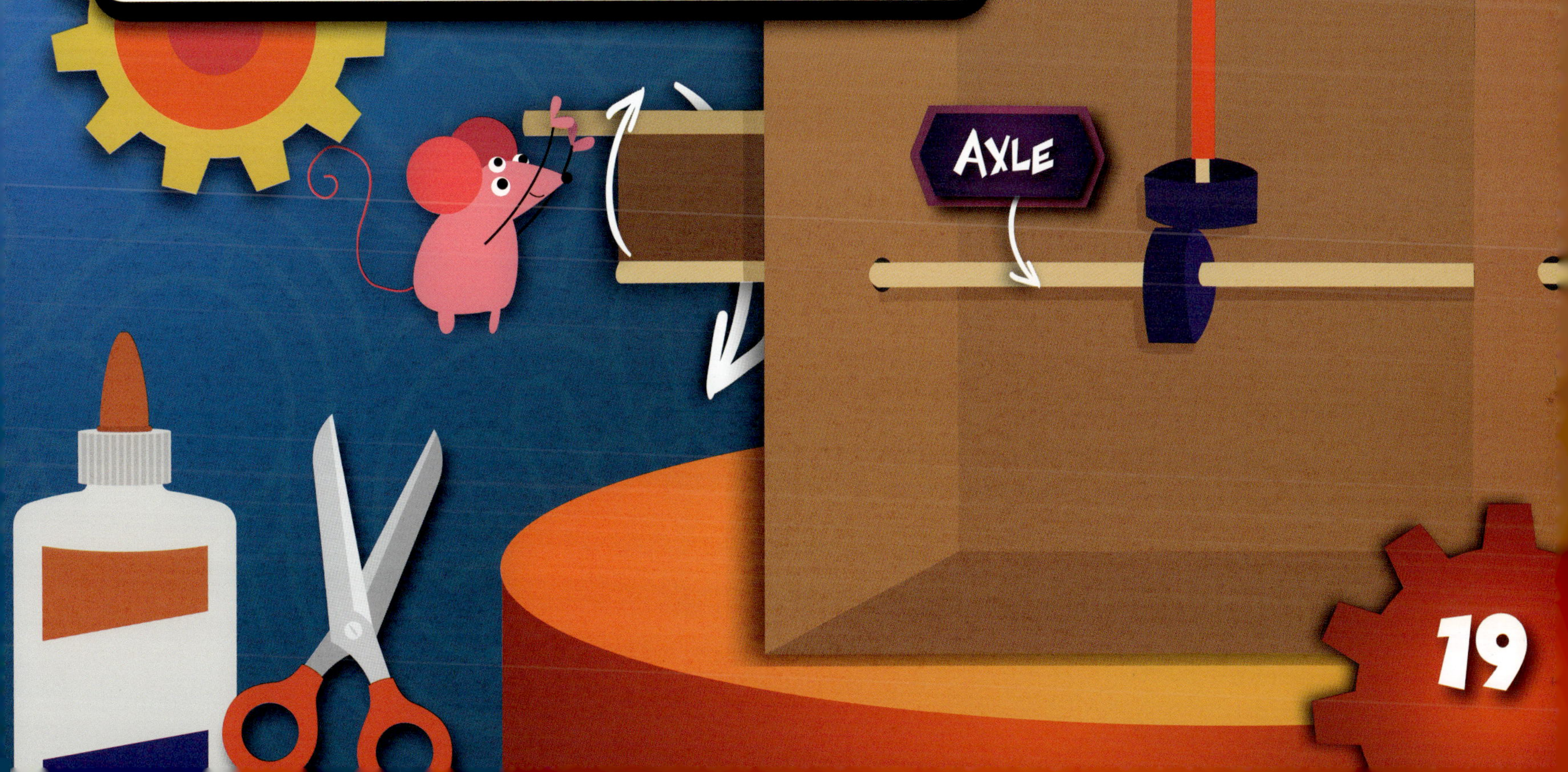

STEP 1

Cut the front off your cardboard box.

STEP 2

Poke a hole in the top of your box, and stick the straw through. Trim the straw, and glue it in place.

STEP 3

Put a stick through the straw, and glue a bottlecap to the end inside the box. Hold it in place while it dries.

STEP 4

Using scissors, poke a hole through the other bottlecap.

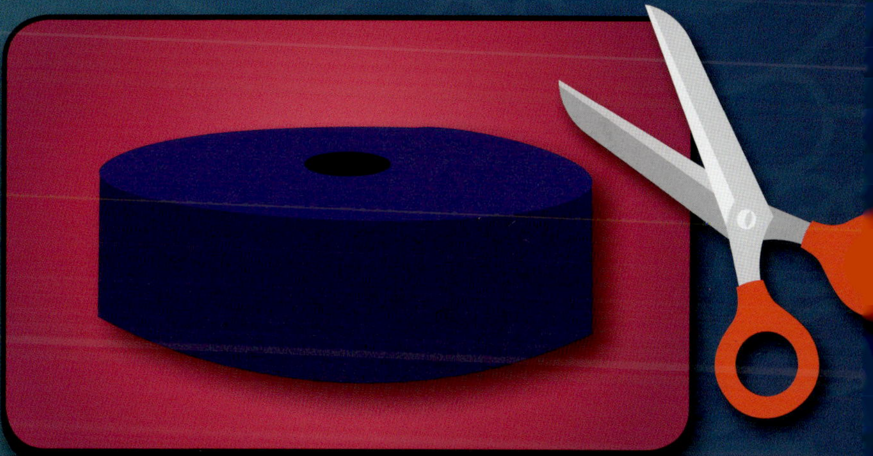

STEP 5

Push a stick through one side of the cardboard box, then through the bottlecap, then through the other side of the box. Keep the stick level, like in the picture.

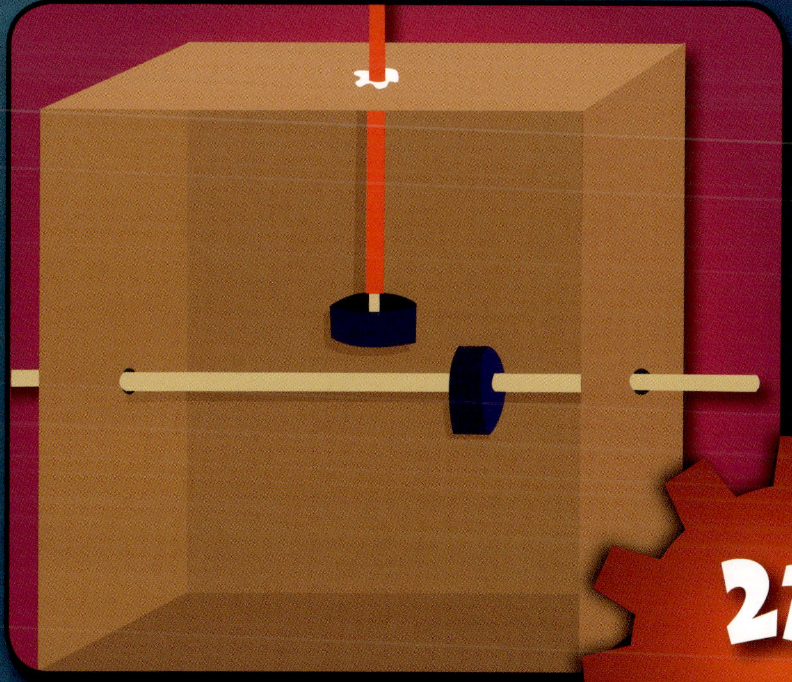

STEP 6

Slide the bottlecap along the stick until it is in the middle, then glue it in place.

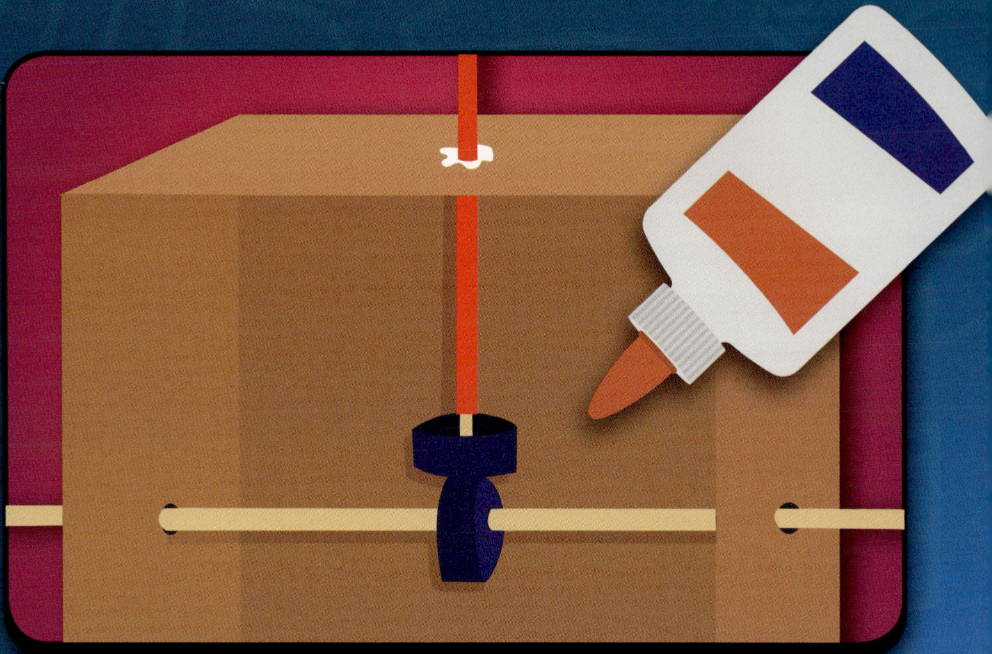

STEP 7

Make a handle for one end. First, glue a very small stick along one side of a square of cardboard. Glue the opposite side of the square to the end of the sideways stick.

STEP 8

Use your colored pencils and other pieces of cardboard to create a character to spin around. It can be whatever you want!

STEP 9

Glue your character to the stick coming out of the top. Line the bottlecaps up like in the picture. Now turn the handle and watch your character spin.

GLOSSARY

electric motor	a machine that moves things using electricity
force	a push or pull on an object
friction	a force that slows things down and creates heat when two things rub together
power	the ability something has to get things done
speed	how fast something goes
vehicle	a machine that has an engine and is used to carry people or things

INDEX

bearings, 11
bikes, 6, 10, 17
cars, 6
electricity, 10
friction, 11
gears, 13–17
machines, 4–5, 10, 13–14, 16
pedaling, 10, 17
speed, 12–13, 16–17
spinning, 6–7, 11–12, 15, 17–18, 23
wheels, 6–14, 17